NATIONAL GEOGRAPHIC KiDS

美国国家地理
双语阅读

U0179965

Planets

行星

懿海文化 编著

马鸣 译

第三级

外语教学与研究出版社
FOREIGN LANGUAGE TEACHING AND RESEARCH PRESS
北京 BEIJING

京权图字：01-2021-5130

图书在版编目（CIP）数据

行星：英文、汉文 / 懿海文化编著；马鸣译. —— 北京：外语教学与研究出版社，2021.11（2023.8 重印）

（美国国家地理双语阅读. 第三级）

书名原文：Planets

ISBN 978-7-5213-3147-9

I. ①行⋯ II. ①懿⋯ ②马⋯ III. ①行星－少儿读物－英、汉 IV. ①P185-49

中国版本图书馆 CIP 数据核字 (2021) 第 236727 号

出 版 人　王　芳
策划编辑　许海峰　刘秀玲　姚　璐
责任编辑　姚　璐
责任校对　华　蕾
装帧设计　许　岚
出版发行　外语教学与研究出版社
社　　址　北京市西三环北路 19 号（100089）
网　　址　https://www.fltrp.com
印　　刷　天津海顺印业包装有限公司
开　　本　650×980　1/16
印　　张　37.5
版　　次　2022 年 3 月第 1 版　2023 年 8 月第 4 次印刷
书　　号　ISBN 978-7-5213-3147-9
定　　价　188.00 元（全 15 册）

如有图书采购需求，图书内容或印刷装订等问题，侵权、盗版书籍等线索，请拨打以下电话或关注官方服务号：
客服电话：400 898 7008
官方服务号：微信搜索并关注公众号"外研社官方服务号"
外研社购书网址：https://fltrp.tmall.com

物料号：331470001

记载人类文明
沟通世界文化
www.fltrp.com

Table of Contents

What's a Planet?

Circle, circle in the sky,
you're bright enough
to catch my eye.
You're not a star,
but a place
where gas or rocks
swirl in space.

Space Clues

GAS: Something that
has no shape or size of
its own. Gas can spread
out into the space
around it.

STAR: A huge ball of
hot gas that makes heat
and light

Hello, Planets!

When you look up high in the night sky, you might see lots of bright lights. Most of these lights are stars.

Space Clues

ORBIT: The path an object follows around another object, such as a star

REFLECT: To bounce back

Q What did two stars say to each other on Valentine's Day?

A I glow for you!

Most stars have planets moving around them. Planets are round objects that orbit a star. They don't create their own light. They only reflect light from stars.

The Sun

The sun is our star. Our planet Earth orbits the sun. The sun makes a lot of heat and light.

The sun

Don't look directly at the sun. It could hurt your eyes.

Its surface is about 100 times hotter than a hot summer day! It's also very big. One million Earths could fit inside the sun!

Earth

Our Solar System

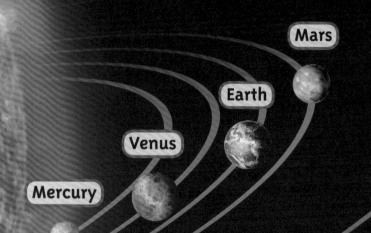

Mars

Earth

Venus

Mercury

Our sun and Earth are part of what we call our solar system. There are eight big planets and five small, dwarf planets in the solar system.

Neptune

Uranus

Saturn

Jupiter

The sun and the eight big
planets in our solar system

Each planet orbits the sun. The
strong pull of the sun's gravity
holds the planets in orbit, which
keeps them from floating
away. Gravity is the
same force that makes
a baseball fall to the
ground when you
drop it.

Space Clue

GRAVITY: An
invisible force
that pulls objects
toward a planet
or star

The Inner Planets

Mercury, Venus, Earth, and Mars are in the hotter part of our solar system.

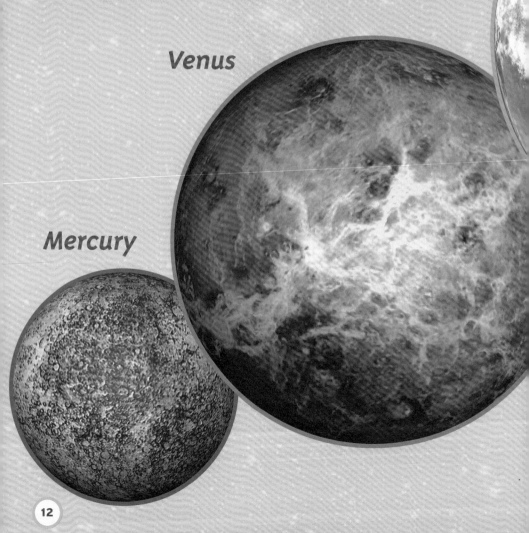

Venus

Mercury

Mars

Earth

These planets are the four closest
to the sun. They are made of rock
and metal.

Your Planet Earth

Earth is the third planet from the sun. The planet we call home spins at just the right distance from the sun.

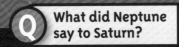
It's far enough away to be not too hot, yet close enough so it's not too cold. Here, the conditions are just right for life.

sea animals

plants

land animals

15

The Gas Giants

Saturn

rings

The gas giants do not have a hard surface like the rocky inner planets.

Neptune

Jupiter

Uranus

Saturn

Beyond the rocky inner planets, there are four gas giants. They are Jupiter, Saturn, Uranus (YOOR-eh-nes), and Neptune. These huge planets are made of big clouds of gas and liquid. Gravity pulls the gas and liquid into a planet shape.

All gas giants have rings. The rings are made of mostly ice and dust.

Dwarf Planets

Dwarf planets are planetlike objects that are part of our solar system. They are much smaller than regular planets.

MakeMake (MAH-keh-MAH-keh)

Ceres (SEER-eez)

Pluto

Eris (EHR-is)

This dwarf planet is shaped like an egg.

Haumea
(Ha-oo-MAY-ah)

A dwarf planet can be round or egg-shaped. Unlike regular planets, dwarf planets may share their orbits with other space objects.

Amazing Planets

These planets are really out of this world!

WEIRDEST SPIN

Uranus

This planet spins on its side, rolling like a barrel instead of a top.

WINDIEST WEATHER

Neptune

Winds on Neptune blow much faster than Earth's strongest hurricanes.

LONGEST-LASTING STORM

Jupiter

The Great Red Spot is a hurricane on Jupiter. It has been blowing for more than 400 years.

Great Red Spot

HOTTEST PLANET

Venus

This is the solar system's hottest planet even though it is not the closest to the sun. A thick layer of gas around Venus makes it super hot.

TALLEST MOUNTAIN

Mars

Mars is the home of the solar system's largest volcano, called Olympus Mons. It's as tall as three Mount Qomolangmas stacked up.

Olympus Mons

Mount Qomolangma

Moons Galore!

Some planets, like Earth, have moons that travel with them. Moons are objects made of ice or rock that orbit a planet.

Some planets have no moons. Some have many. Jupiter has more than 60!

Jupiter and four of its moons

Jupiter

Saturn has a giant moon called Titan. It is one of the largest moons in the solar system. It is even bigger than the planet Mercury!

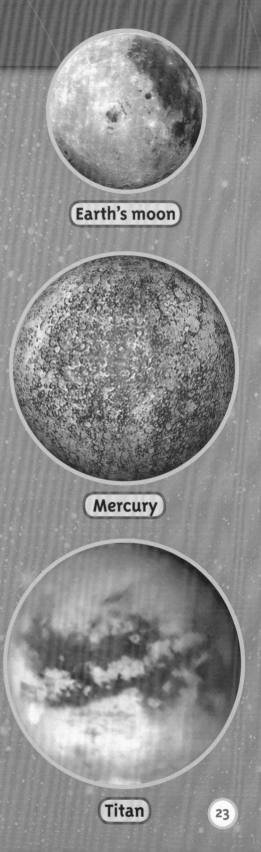

Earth's moon

Mercury

Titan

Our Moon

Earth has only one moon. It appears large and bright in the night sky.

Astronauts first walked on the moon more than 40 years ago. They left footprints in the dirt. There is no weather on the moon to wash or blow them away. Those footprints are still there.

Space Clue

WEATHER: The changing conditions that can include temperature, rainfall, wind, and clouds

Astronaut Buzz Aldrin walks on the moon in 1969 as part of the Apollo 11 mission.

Exploring Space

Scientists have many ways of studying planets. One is to visit them. Humans can't visit other planets yet. But we can send robots to explore them for us.

A rocket blasts off from Cape Canaveral, Florida. It is heading to Mars.

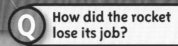

Robots called rovers have been sent to Mars. The rovers carry cameras and tools. They can take photographs and videos, too. The rovers send information back to Earth about what they see.

A drawing of a rover on Mars

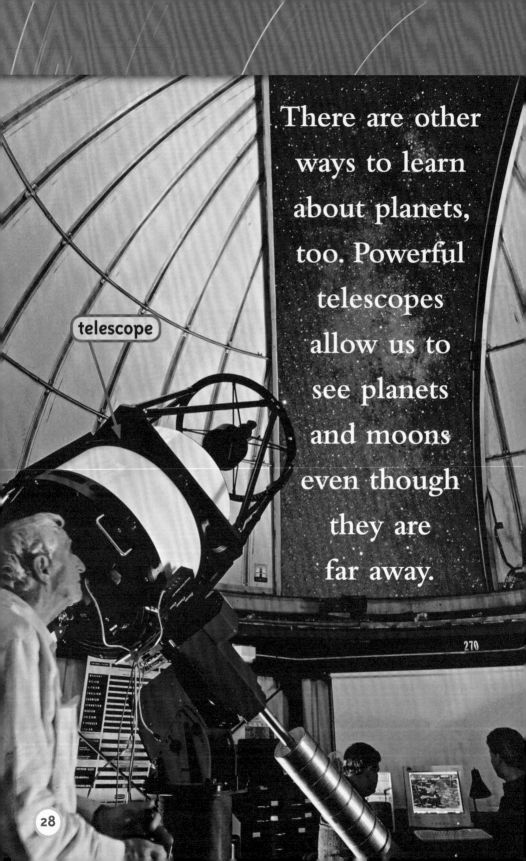

There are other ways to learn about planets, too. Powerful telescopes allow us to see planets and moons even though they are far away.

telescope

A drawing of a space probe that flew near Jupiter

Space probes also collect information about planets. These spacecraft take pictures. They measure light and temperature, too.

With these tools, scientists make exciting discoveries about other worlds. What will we learn next?

Stump Your Parents

Can your parents answer these questions about planets? You might know more than they do!

Answers are at the bottom of page 31.

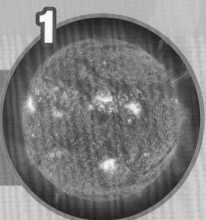

Which is NOT true about the sun?

A. It makes heat and light.
B. It is our planet's star.
C. It is much smaller than Earth.
D. Planets orbit around it.

What does Neptune have that's so special?

A. The largest volcano
B. The hottest surface
C. The weirdest spin
D. The windiest weather

What is Jupiter's Great Red Spot?

A. A moon
B. A volcano
C. A flood
D. A hurricane

4

Which planet has just one moon?

A. Mercury
B. Earth
C. Mars
D. Uranus

5

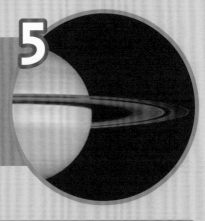

Which of these are the main materials found in planets' rings?

A. Gas
B. Ice
C. Dust
D. B and C

6

What can you use to get a better view of objects in space?

A. Microscopes
B. Telescopes
C. Submarines
D. Flashlights

7

To study the environment on Mars, scientists have sent ____.

A. Rovers
B. Airplanes
C. Space shuttles
D. Astronauts

Glossary

GAS: Something that has no shape or size of its own. Gas can spread out into the space around it.

GRAVITY: An invisible force that pulls objects toward a planet or star

ORBIT: The path an object follows around another object, such as a star

REFLECT: To bounce back

STAR: A huge ball of hot gas that makes heat and light

WEATHER: The changing conditions that can include temperature, rainfall, wind, and clouds

▶ 第 4—5 页

什么是行星?

你在天空中转呀转,明亮闪耀,夺我眼球。你不是恒星,而是太空中气体或岩石旋动之所在。

行星小词典

气体:一种本身没有形状或大小的东西。气体可以扩散到周围的空间。

恒星:炽热的气体形成的巨大球体,会发光、发热

▶ 第 6—7 页

你好,行星!

行星小词典

轨道:一个天体环绕另一个天体——比如恒星——运行的路径

反射:到达一个表面后快速返回

当你抬头遥望夜空,你会看到很多亮光。这些亮光大都是恒星。

大多数恒星都有行星环绕其运行。行星是在轨道上环绕恒星运行的球状天体。它们本身不发光。它们只反射恒星的光。

▶ 第 8—9 页

太阳

太阳是恒星。行星地球在轨道上环绕太阳运行。太阳发出大量的光和热。它的表面比炎热的夏天还要热大约 100 倍!它也非常大。太阳可以容纳 100 万个地球!

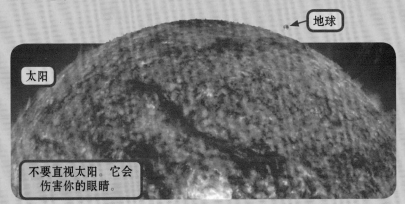

地球

太阳

不要直视太阳。它会伤害你的眼睛。

▶ 第 10—11 页

太阳系

海王星
天王星
土星
木星
火星
地球
金星
水星

太阳系中的太阳与
八大行星

　　太阳和地球是太阳系的一部分。太阳系有八颗大行星和五颗矮行星。
　　每颗行星都在轨道上环绕太阳运行。太阳巨大的引力使行星一直待在轨道上，不会偏离。引力就是使你投掷出去的棒球落在地上的那种力。

行星小词典

引力：一种看不见的、能把物体拉向行星或恒星的力

▶ 第 12—13 页

内行星

　　水星、金星、地球和火星在太阳系中比较热的部分。
　　这些行星是最靠近太阳的四颗行星。它们是由岩石和金属构成的。

火星

金星

地球

水星

▶ 第 14—15 页

行星地球

地球是距离太阳第三近的行星。这颗我们称之为"家园"的行星在距离太阳适当的位置旋转。

这个距离足够远，不会太热，也足够近，不会太冷。这里的条件刚好适合生命生存。

植物

海生动物

陆生动物

▶ 第 16—17 页

气态巨行星

海王星

木星

天王星

土星

土星

光环

气态巨行星没有岩质内行星那样的坚硬表层。

除了岩质内行星，还有四颗气态巨行星，它们是木星、土星、天王星和海王星。这些巨大的行星是由大气团和液体构成的。引力使气体和液体汇聚成行星的形状。

所有的气态巨行星都有光环。这些光环主要由冰和尘埃构成。

▶ 第 18—19 页

矮行星

矮行星是像行星一样的天体，是太阳系的一部分。它们比一般的行星小得多。

鸟神星

谷神星

冥王星

阋神星

这颗矮行星很像一枚蛋。

矮行星有球形的，也有蛋形的。跟一般的行星不同，矮行星可与别的天体共享轨道。

神奇的行星

这些行星真的很独特!

最奇怪的旋转

天王星

这颗行星侧着旋转,像桶那样而不像陀螺那样滚动。

风力最强的天气

海王星

海王星上的风刮得比地球上最强的飓风还要快。

最持久的风暴

木星

大红斑

大红斑是木星上的飓风。它已经刮了超过400年。

最热的行星

金星

虽然它不是离太阳最近的行星,但它是太阳系最热的行星。厚厚的大气层环绕着金星,这使得它非常热。

最高的山峰

火星

火星是太阳系最大的火山——奥林匹斯山的所在地。奥林匹斯山有三座珠穆朗玛峰叠起来那么高。

奥林匹斯山

珠穆朗玛峰

▶ 第 22—23 页

好多卫星！

有些行星，比如地球，有卫星和它们一起运行。卫星是由冰或岩石构成的天体，在轨道上环绕行星运行。

一些行星没有卫星。一些行星有很多卫星。木星有超过60颗卫星！

土星有一颗巨大的卫星，叫"土卫六"。它是太阳系最大的卫星之一。它甚至比行星水星还要大！

地球的卫星

木星和它的四颗卫星

木星

水星

土卫六

▶ 第 24—25 页

卫星月球

地球只有一颗卫星。它在夜空中又大又亮。

四十多年前，宇航员第一次在月球上行走。他们在月球表面的尘埃上留下了足迹。月球上没有任何天气变化可以将尘埃冲走或吹走。那些足迹仍然在那儿。

行星小词典

天气：包括温度、降水、风和云等在内的变化状况

1969年，宇航员巴兹·奥尔德林在月球上行走，这是"阿波罗11"号的任务之一。

▶ 第 26—27 页

探索太空

　　科学家有很多种方式研究行星，其中一种是造访它们。人类目前还不能造访其他行星。但是我们可以发射机器人为我们探索行星。

一枚火箭在佛罗里达州的卡纳维拉尔角市发射升空。它飞往火星。

　　一种被称为"火星漫游车"的机器人已经被送到了火星上。火星漫游车携带照相机和工具。它们可以拍照、录影。火星漫游车把观测到的信息传回地球。

火星漫游车在火星上的画面

▶ 第 28—29 页

　　还有其他方式了解行星。高倍望远镜可以让我们看到行星和卫星——尽管它们离我们很远。

望远镜

在木星附近飞行的航天探测器的画面

　　航天探测器也可以收集有关行星的信息。这些航天器能拍照。它们也能测量光和温度。

　　借助这些工具，科学家对地球以外的世界有了很多令人激动的发现。接下来我们将了解什么呢？

挑战爸爸妈妈

你的爸爸妈妈能回答这些有关行星的问题吗？你可能比他们知道的还多呢！答案在第 31 页下方。

1 关于太阳的说法哪一个不是真的？
A. 它发光，发热。　　B. 它是行星的恒星。
C. 它比地球小得多。　D. 行星在轨道上环绕它运行。

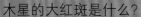

2 海王星有什么特别之处呢？
A. 最大的火山　　　　B. 最热的表层
C. 最奇怪的旋转　　　D. 风力最强的天气

3 木星的大红斑是什么？
A. 一颗卫星　　　　　B. 一座火山
C. 一场洪水　　　　　D. 一场飓风

4 哪颗行星只有一颗卫星？
A. 水星　　　　　　　B. 地球
C. 火星　　　　　　　D. 天王星

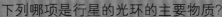

5 下列哪项是行星的光环的主要物质？
A. 气体　　　　　　　B. 冰
C. 尘埃　　　　　　　D. B 和 C

6 你可以用什么更好地观察天体？
A. 显微镜　　　　　　B. 望远镜
C. 潜水艇　　　　　　D. 手电筒

7 为了研究火星上的环境，科学家发射了 _____。
A. 火星漫游车　　　　B. 飞机
C. 航天飞机　　　　　D. 宇航员

▶ 第32页

词汇表

气体：一种本身没有形状或大小的东西。气体可以扩散到周围的空间。

引力：一种看不见的、能把物体拉向行星或恒星的力

轨道：一个天体环绕另一个天体——比如恒星——运行的路径

反射：到达一个表面后快速返回

恒星：炽热的气体形成的巨大球体，会发光、发热

天气：包括温度、降水、风和云等在内的变化状况